i-SPY

fossils
and rocks
SPY IT! SCORE IT!

Introduction

Are you a keen geologist? This book is for you if you're interested in fossils and rocks, live near to an area known for them, or are planning to visit somewhere with a collection of fossils or rocks.

You can learn how to identify them, and the most common locations in the UK to seek them out – you may spot them in other places not mentioned here. There is a handy glossary at the back of the book to help you too.

So, keep a sharp look-out for fossils and rocks, and get spotting!

How to use your i-SPY book

Keep your eyes peeled for the i-SPYs in the book.

If you spy it, score it by ticking the circle or star.

Items with a star are difficult to spot so you'll have to search high and low to find them.

50 POINTS

Once you score 1000 points, send away for your super i-SPY certificate. Follow the instructions on page 64 to find out how.

Tools

Here are some of the things you may need when you're spotting fossils and rocks.

Hi-vis waistcoat

Waterproofs

Hard hat

Notebook and pen

Rucksack

Measuring tape

Safety glasses

Geological hammer

Waterproof boots

Collecting bag

Rock-handling gloves

Magnifying lens

Minerals

Minerals are solid, inorganic chemical elements or compounds that are formed naturally in the ground.

Quartz

Natural form Crystals are often long, showing six large faces parallel to the length of the crystal; massive in veins

Uses Quartz sand in the building industry, abrasives, semi-precious stones

Colour Usually colourless or white; semi-precious varieties are coloured, for example, rose quartz, citrine (yellow), smoky quartz, amethyst (mauve)

Where to spot it Abundant in many rocks; widespread

10 POINTS

Orthoclase feldspar

Natural form Crystals are stumpy, and look like rectangles in a rock

Uses For making porcelain and its surface glass

Colour White to pale pink; white streak; pearly to glassy lustre

Where to spot it Abundant in granites and many other igneous and metamorphic rocks

20 POINTS

4

Plagioclase feldspar

Natural form Elongate or tabular crystals, or massive

Colour White, less commonly pink or with green or brown; the variety labradorite looks bluish or grey

Where to spot it Abundant in many igneous rocks; widespread

Uses In the ceramics industry

Micas

Natural form Dark mica (biotite) or light mica (muscovite); flexible flakes

Colour Muscovite: colourless, green, or pale brown. Biotite: dark brown, greenish-black or black

Where to spot it In many igneous and metamorphic rocks; widespread

Uses In the electrical and electronics industries, as a filler in paints and plastics

Augite

Natural form Massive or granular, or as crystals that are eight sided in cross-section

Colour Dark green, black; white streak; glassy lustre

Where to spot it In igneous rocks that have a low quartz content; difficult to distinguish from hornblende

Uses An important rock-forming mineral in igneous and metamorphic rocks. Perfect crystals are sought by collectors

30 POINTS

Hornblende

Natural form Long or short crystals with a six-sided cross-section, or as granular or fibrous masses

Colour Shades of green to almost black; glassy lustre

Where to spot it In metamorphic rocks that have formed through changes in temperature and pressure; also in igneous rocks such as syenite and diorite; widespread

Uses An important rock-forming mineral in igneous and metamorphic rocks. Perfect crystals are sought by collectors

30 POINTS

Olivine

Natural form As grains in basalts or as poorly shaped masses

Colour Olive green, white, yellow-brown, or black

Where to spot it In rocks low in quartz, such as basalt and gabbro, and in some marbles; widespread

Uses Sometimes used in blast furnaces to remove impurities from metal; semi-precious stone (peridot)

40 POINTS

Calcite

10 POINTS

Natural form Calcite can show many different shapes. When the crystals look like a row of dog's teeth, it is called dog-tooth spar. Rhombohedral shapes are common

Colour Colourless or white if pure, but can be almost any other colour if not; white streak; glassy lustre

Where to spot it An important mineral in limestones and marbles; as a cement in sedimentary rocks; also found in veins; widespread

Uses One of the most widely used minerals; in the construction industry for aggregate and cement production

Pyrite

Natural form Cubic crystals

Colour Brassy or golden yellow; black streak; metallic lustre

Where to spot it In black shales as cubes and nodules; sometimes replaces fossils such as ammonites; may occur in mineral veins or as small grains in igneous rocks; widespread

Uses In the manufacture of sulphuric acid; ornamental

10 POINTS

Gypsum

Natural form Tabular crystals with curved faces; fibrous, massive, or granular; may grow in rosettes to form what is called a desert rose

Colour Colourless, white or possibly yellow, grey or brown due to impurities; white streak; glassy, pearly or dull lustre

Where to spot it Found today in areas where sea water is evaporating, or as sedimentary rock formed in the same environment; widespread

Uses In the manufacture of cement, plaster of Paris and plasterboard; for agricultural purposes, and for plaster for surgical splints

20 POINTS

Fluorite

Natural form
Crystals are cubic (with six faces) or octahedral (with eight faces)

Colour Varied, including colourless, purple, green, yellow, blue or pink; white streak; glassy lustre

Where to spot it
In mineral veins in limestone areas associated with quartz, baryte, and galena, such as the Pennines and the Lake District

Uses Ornamental; in smelting and in chemical industries, for example, to produce hydrofluoric acid

30 POINTS

Garnet

Natural form
Crystals common with rhombic faces; massive

Colour Varies with type, for example, almandine is red-brown to black, uvarovite is bright green; glassy or resinous lustre

Where to spot it
In schists, gneisses, serpentinites, granites and pegmatites, such as in the Lake District and Scotland

Uses In abrasives and as gemstones

20 POINTS

Galena

Natural form
Cubic; may also be massive or granular

Colour Dull lead grey; dull grey streak; metallic lustre

Where to spot it
In veins, often with sphalerite, and in sedimentary rocks that have been soaked with hot fluids rising from deep inside the Earth, in areas such as southwest England, the Pennines and Scotland

Uses An important lead ore; may also contain small amounts of silver

30 POINTS

Hematite

Natural form Tabular crystals, rose-like growths and domed masses (kidney ore)

Colour Reddish-black, steel grey, black; dark-red streak; metallic or dull lustre

Where to spot it Usually in sandstones and limestones affected by mineral-rich fluids; widespread; occurs on Mars

Uses An important iron ore; used in stains and pigments

30 POINTS

Sphalerite

Natural form Crystals are tetrahedral (with four faces), or as complex variations. Also massive, part-crystalline or even fibrous

Colour Resinous yellow through various shades of translucent or even opaque brown

Where to spot it As veins and deposits, often with galena, fluorite and other minerals, e.g. where hot fluids have flowed into limestone or other sedimentary rock

Uses The most valuable ore of zinc

40 POINTS

TOP SPOT!

Magnetite

Natural form Crystals octahedral (with eight faces); also massive or granular

Colour Black; silvery-black due to reflections from surface

Where to spot it Common as small grains in all igneous rocks and metamorphic rocks; also in beach sand

Uses Because it is strongly magnetic, it was used as a compass (lodestone) in early times. Of economic value as iron ore

50 POINTS

Azurite and Malachite

Natural form Azurite crystals are very rare; the mineral occurs more often as radiating clusters and soil-like masses. Malachite can form attractive rounded and banded (striped) masses, or occur as fibres; crystals are rare

Colour Azurite has vivid, azure-blue transparent to translucent crystals and a light-blue streak. Malachite is bright green with a pale-green streak and dull to silky sheen

Where to spot it Often found together in the upper zones of copper deposits, near the Earth's surface, although malachite is the more common

Uses Good specimens are prized by collectors. Both are used in searching for copper and have been used as pigment since ancient times. Polished specimens are very decorative

40 POINTS

Igneous rocks

There are three main types of rock: igneous, sedimentary and metamorphic. Igneous rocks (from the Latin word for 'fire') start out being so hot that they are liquid! As the molten rock (magma) cools, igneous rocks form.

Granite

Grain size Coarse (because it cooled slowly at great depth in the Earth's crust)

Colour White, pink or grey mottled appearance

Texture Well-formed crystals of white feldspar show a rectangular shape and may be larger than the other crystals (porphyritic texture)

Where to spot it Devon and Cornwall, Ben Nevis in Scotland and the Mourne Mountains, County Down

Uses Constructing buildings, bridges, monuments and as rail ballast. Indoors it can be used to make kitchen counter-tops and for flooring

10 POINTS

Granite pegmatite

Grain size Very coarse but variable

Colour Uneven or patchy; mainly white, pink or red

Texture Grains can be 10 cm long or more; the world's biggest crystals are found in pegmatite and can be many metres long

Where to spot it In granite regions such as Scotland and southwest England

Uses May contain gem minerals such as beryl, emerald, garnet, ruby, sapphire and topaz

25 POINTS

Rhyolite

Grain size Very fine-grained or glassy, because of quick cooling at the Earth's surface. The same composition as granite

Colour Light grey, reddish or brown

Texture Fine-grained but may be porphyritic (contain some bigger crystals); may show stripes of colour indicating flow

Where to spot it The Lake District and the Isle of Lundy

Uses As aggregate in the construction industry; as a facing stone; in jewellery

30 POINTS

Syenite

Grain size Coarse

Colour Grey, pink or red; the closely related and attractive larvikite is grey to bluish as it contains a lot of labradorite

Texture Coarse crystals, mostly feldspar

Where to spot it The Highlands and Southern Uplands of Scotland

Uses Decorative stone, polished slabs, road stone. Huge boulders of larvikite are often used for sea defences

40 POINTS

Andesite

Grain size Fine to partly glassy (because it cooled quickly at the Earth's surface)

Colour Normally shades of grey

Texture Fine-grained, but often contains some large crystals (porphyritic texture)

Where to spot it The Lake District, Scotland

Uses Making tiles; road construction; for sculptures and monuments

Score double points for seeing the Ice-Age Andesite Boulder at the University of Manchester.

30 POINTS

Diorite

Grain size Coarse (it is the same composition as andesite, but cooled much more slowly deep in the Earth)

Colour Speckled with grey or white, almost like salt and pepper, sometimes with a little green or pink

Texture Grains usually of equal size

Where to spot it Leicestershire, Aberdeenshire, the Isle of Skye, Perth and Kinross

Uses Used in the construction of roads and buildings; museum sculptures

40 POINTS

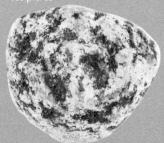

The next three igneous rocks all have the same composition but have different crystal sizes because they cooled down at different rates.

Gabbro

Grain size Coarse (cooled slowly)

Colour Dark grey to black augite, speckled with plagioclase feldspar, which is white or may have a greenish or purply tinge

Texture Coarse crystals of equal size

Where to spot it The Lake District, the Hebrides, South Ayrshire

Uses Crushed stone in building projects; facing stone (for example, on the walls of buildings); gravestones

10 POINTS

Dolerite

Grain size Medium

Colour Dark grey to black, may be mottled

Texture Crystals are generally less than 2 mm; quite smooth to the touch

Where to spot it Often seen as sheets intruded into local rock, particularly in the Hebrides, southern Scotland, northeast England, Antrim, North Wales

Uses Building and decorative stone

10 POINTS

Basalt

Grain size Fine (individual crystals can't be seen)

Colour Dark grey to black

Texture May contain visible crystals of, for example, olivine; sometimes appears ropey or blocky, or may form columns. Often quite smooth to the touch but can contain cavities (holes); known as vesicles, these are sometimes filled with minerals

Where to spot it Southwest England, north Wales, Yorkshire, the Lake District, Scotland, Ireland

Uses Used in construction and for making statues

Look for the almost glassy edges of these rocks, where they cooled quickly against the surrounding rock.

10 POINTS

The Giant's Causeway in Northern Ireland has over 40,000 basalt columns.

Obsidian

Grain size None (it's glassy, but opaque except at the thin edges)

Colour Black, red, brown; the variety snowflake obsidian shows greyish-white 'snowflakes'

Texture Can have streaks or bands of air bubbles, or of minute crystals. Breaks with a concentric fracture; the edges are extremely sharp

Where to spot it In museums – rarely found in the UK

Uses Knives, arrows and spearheads up to the 20th century, and still used for ceremonial weapons in some cultures

Score an extra 10 points for spotting a weapon made from obsidian in a museum.

50 POINTS

TOP SPOT!

Pumice

Grain size Glassy with some fine crystals

Colour Very light grey or yellowish

Texture Rough to the touch because there are lots of cavities (holes) caused by escaping gas as it formed. These vary in size and may be lined up.

Where to spot it Beaches of the Western Islands of Scotland, Orkney and Shetland

Uses In the production of lightweight concrete; cosmetic industry (pumice stone)

Most rocks sink in water. Some types of pumice float because they are less dense than water.

30 POINTS

Tuff

Grain size Fine

Colour Newly fallen ash is white or grey; when solidified into rock called tuff, it can be shades of brown, red or dark green

Texture May contain mixtures of crystals, glass and pieces of rock; can be well-bedded. If it contains pea-sized fragments (small pieces), it is called lapilli tuff

Where to spot it Borrowdale in the Lake District, East Lothian in Scotland

Uses Construction, floors and facing stone

20 POINTS

TOP SPOT!

Porphyry

Big, well-formed crystals in a fine-grained lava. Various colours. Common in 'till' that was carried by ice from Oslo in Norway to the Yorkshire coast during the Ice Age

50 POINTS

Metamorphic rocks

Igneous or sedimentary rocks that have had their original structure or form changed by pressure and/or heat are known as metamorphic rocks.

10 POINTS

Marble

Grain size Medium to coarse

Colour White when pure; grey, pink and sometimes black

Texture Interlocking crystals, and can look like sugar; may show stripes

Where to spot it The islands of Iona, Tiree and Skye in Scotland

Uses Jewellery, building stone, decorative stone, gravestones; often used by sculptors

Marble is altered limestone, often discoloured by green olivine or other coloured minerals

Serpentinite

Grain size Medium to coarse

Colour From yellow-green to black, often with bright green or red streaks and blotches

Texture Feels greasy or waxy; massive or with lots of polished shear surfaces which can't be separated easily

Where to spot it Cornwall, Scotland

Uses Ornaments, facing stone (on the walls of buildings)

40 POINTS

Slate

Grain size Fine; too small to see

Colour Grey to black; can be purple or greenish

Texture Splits into thin sheets, giving it a layered appearance; smooth to the touch; sometimes shows oval green spots; originally formed from muddy rock, slate may still show some fossils, but they will have been squashed into a different shape

Where to spot it Southwest England, north Wales, the Lake District

Uses Floor tiles, roofing material, billiard tables, gravestones

5 POINTS

Mica Schist

Grain size Medium to coarse; crystals large enough to see

Colour Greyish and sparkling from micas, but white quartz grains may be seen

Texture Flaky; can split into thin sheets, giving it a layered appearance; may contain larger crystals of newly formed minerals such as garnet (called porphyroblasts)

Where to spot it Widespread, for example, southwest England, north Wales, Scotland

Uses Schist is not very strong, so has not been used much for building. It sometimes contains minerals of gem quality

25 POINTS

Gneiss

Grain size Medium to coarse

Colour Multi-coloured with stripes of grey, pink, white or black

Texture Stripes and streaks of light and dark minerals. One variety shows large 'eye-shaped' crystals of pink feldspar, known as augen. May contain large crystals (porphyroblasts) of new minerals such as garnet

Where to spot it Scotland

Uses Roadstone, decorative stone

 10 POINTS

Hornfels

Grain size Fine

Colour Dark grey, brown, greenish, reddish

Texture Crystals generally small and all the same size. May contain crystals of new minerals, such as needle-like andalusite. The rock is very tough and hard to break

Where to spot it Forms next to igneous rocks such as granite, for example, in the Lake District, the Isle of Skye and the southern Highlands of Scotland

Uses Flooring, paving, decorative stone

 20 POINTS

Quartzite

Grain size Medium

Colour Often white, pink or grey but can be a variety of colours

Texture Hard and sugary, with interlocking crystals

Where to spot it Throughout the UK and Ireland, including Shropshire, Nuneaton, Sperrin Mountains, Northern Ireland, Shetland Islands, Scotland

Uses Road building

Quartzite is a metamorphic rock formed from quartz sandstone.

5 POINTS

Eclogite

Grain size Medium to coarse

Colour A beautiful red and green rock; the red colour is from garnet and the green is from a mineral similar to augite

Texture Often has a banded structure, but can be massive; garnet crystals can be very large; it is a very dense rock (heavy for its size)

Where to spot it Scottish Highlands. This is a very rare rock, having formed at very high pressures very deep in the Earth, at least 45 km below the surface

Uses Scientists are interested in eclogite because it gives information about what is happening deep in the Earth

Occasionally, people have found diamond in eclogite! Diamonds are made of carbon and form 150 km or more below the Earth's surface, at very high pressure.

50 POINTS

TOP SPOT!

Sedimentary rocks

Sedimentary rocks are formed from sediment (solid material) left by water, ice or wind.

Conglomerate

Grain size Coarse; rounded pebbles

Colour Varies

Texture Particles you can see, from pebbles to rock pieces; rough to the touch

Where to spot it Widespread, for example, Cornwall and the east coast of Scotland

Uses Building stone, decorative stone

In 2012, a NASA robot on Mars discovered what seems to be conglomerate!

10 POINTS

Breccia

Grain size Coarse

20 POINTS

Colour Varies

Texture The same as conglomerate, but with more angular particles

Where to spot it Widespread, for example, Cornwall, Devon, the Midlands, Scotland

Uses Road stone, decorative stone

Sandstone

Grain size From fine to grit

Colour Can be any colour, but grey, green and light brown to red are common

Texture Massive or in thin layers of different grain sizes

Where to spot it A range of places, including Dorset, Wiltshire, Hampshire, southwest Wales, northwest Highlands

Uses Building industry, manufacture of concrete

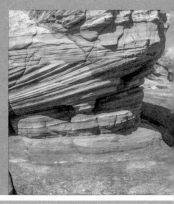

5 POINTS

Mudstone

Very fine-grained sediment, made up of clay and some silt, from soft to hard. If it splits easily into thin layers, it is called Shale.

5 POINTS

Till

Moving glaciers scrape up everything before them and when they melt, they dump it. This mix of mud and boulders is called till and a blanket of it covers much of northern Britain. Look for it along the Yorkshire coast.

25 POINTS

Shelly limestone

Grain size Coarse shell fragments (small pieces)

Colour White, grey, tan, blue

Texture Rough to the touch

Unlike sand and mud, which are made from particles of older, broken-up rocks, most limestones are made of organic material (from animals and plants) which are all calcium carbonate (lime).

Where to spot it A range of places, including Gloucestershire and Yorkshire

Uses Building stone, decorative stone

10 POINTS

Oolitic limestone

You can test if a rock is limestone – it fizzes in vinegar.

Grain size Can reach 2 mm in diameter but mainly around 1 mm

Colour White, yellow-brown

Texture Rounded spheres (like a ball); each sphere is built up in layers

Where to spot it A range of places, including the Cotswold hills, Gloucestershire and Yorkshire

Uses Building stone

20 POINTS

Chalk

Natural chalk was first used for drawing more than 10,000 years ago!

Grain size Fine to very fine

Colour White or grey

Texture Crumbly to hard

Where to spot it North and South Downs, Chilterns, Salisbury Plain, Lincolnshire and Yorkshire Wolds

Uses Road fill, lime, cement manufacture

5 POINTS

Limestone pavement

TOP SPOT!

Bare, weathered limestone pavements can be found in the Pennines and Ireland. They are made of limestone blocks (clints) separated by eroded cracks (grykes).

50 POINTS

Flint

Colour Black to grey (although often stained brown)

Texture Hard, shiny and splintery with a dull cortex (outer layer)

Where to spot it Anywhere that chalk is exposed; as flint gravel in rivers, for example, the Thames

Uses Important in knapping (making) Stone-Age tools

Score 20 points for spying a Stone-Age flint tool.

5 POINTS

Coal

Coal is a light, black organic rock made mainly of carbon. It formed as layers of squashed fossil vegetation interbedded with mudstone and sandstone. Most coal in Britain was formed from huge forests which grew during the Carboniferous period

Uses Coal was widely mined for use as a heat source (when burned) for industry and homes

Where to spot it Traditional coal-mining areas of South Wales, England and southern Scotland

 10 POINTS

Salt

Beds of rock salt formed from evaporated sea water, leaving the salt behind as a clear-to-pink-coloured crystalline rock

Uses Chemical industry; de-icing roads

Where to spot it Salt is mined in Cheshire and Yorkshire but is often seen in heaps at the side of the road during the winter

 30 POINTS

Septarian concretion

Colour Grey to brown

Appearance Hard, round to disc-shaped, formed in mudstone. Shrinkage cracks filled with mineral crystals

Where to spot it South-East England, Yorkshire and Dorset

30 POINTS

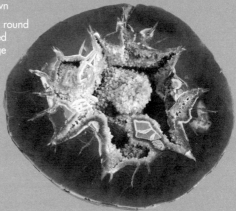

Mudcracks

This rock formed when soft mud on the floor of an ancient lake dried up and became cracked. It acted as a cast for the sand that then flowed over it and became solid rock.

40 POINTS

Structures

Folding

When it is buried deep in the ground, hard rock can become plastic (soft and bendable), and the layers can be folded by pressure within the Earth's crust. Look for Λ-shaped anticlines and V-shaped synclines.

40 POINTS

Faulting

When rock is put under pressure, it can crack and move by a different amount compared with the rock around it, forming a fault. This normally happens deep in the Earth's crust and causes an earthquake at the surface.

40 POINTS

Unconformity

When layers of rock are tilted and eroded (worn down) and then more rock is laid on top, the boundary between them is called an unconformity. When uncomformities were first discovered, they helped to show that the same earth processes have been happening for millions of years.

The De La Beche Unconformity at Vallis Vale in Somerset (shown in the picture) is a good example, as is Hutton's Unconformity at Siccar Point, East Lothian. Score for any unconformity you see.

50 POINTS

TOP SPOT!

Fossils and geological time

The word 'fossil' comes from the Latin 'fossilis', which means 'dug up'. Fossils take thousands or millions of years to form and are the preserved remains of plants, creatures or other lifeforms that have died, or the impressions they leave behind in the sediments around them.

Fossils can be like coins in archaeology – they show the age of the rock in which they are found. This table shows how the millions of years that have passed since the Earth cooled are divided into eras and periods of geological time. The first easily recognised fossils appeared about 570 million years ago, at the beginning of the Cambrian period. To discover where the rocks of different ages are visible at the surface, look at a geological map.

Fossils may be preserved in many ways including: original shell (top), iron pyrite replacement (left), moulds (right).

Era	Period	Epoch/Age	Million years ago	Events
Cenozoic *Age of Mammals*	Quaternary	*Holocene*	*Today*	Ice Age ends Humans are dominant
			0.01	
		Pleistocene		Earliest humans appear Ice Age begins
			1.6	
	Neogene	*Pliocene*		Hominids (human ancestors) appear
			5.3	
		Miocene		Deserts, tundra and grasslands expand Earth looks closer to present day
			23.7	
	Paleogene	*Oligocene*		Mammals are dominant
			36.6	
		Eocene		Mammals, birds and plants diversify
			57.8	
		Paleocene		Tropical forests are widespread
			65.5	
Mesozoic *Age of Reptiles*	Cretaceous			Mass extinction of dinosaurs and other life Flowering plants appear
			144	
	Jurassic			First birds appear Pangaea splits into smaller continents Dinosaurs are dominant
			208	
	Triassic			First dinosaurs appear First mammals appear Reptiles are dominant
			245	
Paleozoic	Permian			Mass extinction of trilobites and other life The supercontinent Pangaea forms
			286	
	Carboniferous	*Age of Amphibians*		First reptiles appear Vast swamp forests and coal beds develop
			360	
	Devonian	*Age of Fishes*		First land animals and amphibians appear
			408	
	Silurian			First land plants appear Fishes dominant First insects appear
			438	
	Ordovician	*Age of Invertebrates*		First vertebrates appear
			505	
	Cambrian			Organisms with shells appear First fungi and corals appear Trilobites are dominant
			570	
Precambrian	Proterozoic Eon			Multicellular organisms appear
			2500	
	Achean Eon			One-celled organisms appear
			3800	
	Hadean Eon			Atmosphere, oceans and oldest rocks form
			4600	

33

Plants

Lepidodendron

Age Carboniferous

Parts of the giant *Lepidodendron* tree were given different names before it was realised that they all belonged to the same plant.

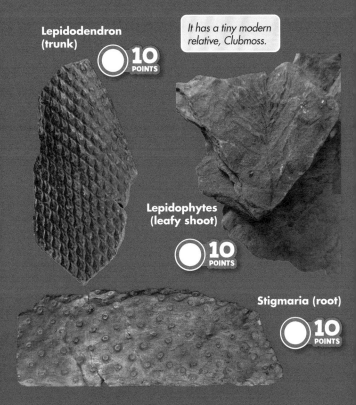

Lepidodendron (trunk)
○ **10** POINTS

It has a tiny modern relative, Clubmoss.

Lepidophytes (leafy shoot)
○ **10** POINTS

Stigmaria (root)
○ **10** POINTS

Calamites

Age Carboniferous

Appearance A bit like ribbed bamboo (it is actually a cast of the central pith cavity)

Where to spot them Northumberland, Wales, Yorkshire, Scotland

This fossil 'horsetail' was once a tree-like plant that grew 20–30 metres high.

10 POINTS

Mariopteris

Age Carboniferous

Appearance A seed fern with lots of detail in the veins of the leaves

Where to spot it Wales, Birmingham, Yorkshire, Scotland

15 POINTS

Coniopteris

Age Jurassic

Appearance Typical fern frond similar to modern examples

Where to spot it In pale grey mudstone blocks on the Yorkshire coast

20 POINTS

Nipa

Age Cenozoic

Appearance Thumbnail-sized, oval-shaped palm tree seed; often looks ribbed

Where to spot it Isle of Sheppey in Kent

30 POINTS

Corals

Corals are made up of small invertebrate animals (animals that don't have a spine), known as zooids. Fossil corals tell us about the past; for example, what the environmental conditions were at certain times in history.

Siphonodendron

Age Lower Carboniferous

10 POINTS

Appearance Classed as a 'Colonial Coral', it has a cluster of long tubes with radiating internal structure

Where to spot it Wales, Scotland, common in pebbles along the Yorkshire coast

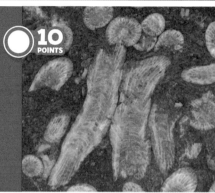

Dibunophyllum

Age Lower Carboniferous

Appearance Classed as a 'Solitary Coral', it is often curved and has easy-to-see internal divisions

Where to spot it Wales, the Midlands, Yorkshire, Scotland

20 POINTS

Sponges

Sponges are simple animals; they don't have muscles, nerves or internal organs like a heart. Fossil sponges are the skeletons of these creatures.

Ventriculites

Age Cretaceous

25 POINTS

Appearance Shaped like a cone; outer part looks like it has grooves or is ribbed

Where to spot it Southeast England, Yorkshire

> *This fossil has been described as looking like a 'petrified mushroom'!*

Raphidonema

Age Lower Cretaceous

Appearance Hollow, like a small lumpy vase

Where to spot it Commonly found all over southern England in paths made from 'Faringdon Sponge Gravel', dug in Oxfordshire

25 POINTS

Brachiopods

Brachiopods live inside a two-part shell. Each part is called a valve. They look similar to cockles and mussels but are not related to them.

Spirifer

Age Late Ordovician to Early Jurassic

Appearance Wing-like appearance with a deep fold down the centre

Where to spot it Southwest England, Wales

Productid

Age Carboniferous to Upper Permian

Appearance Hinged shells of different sizes; the larger shell is domed, the smaller one is flat; both are ribbed

Where to spot it Southwest England, Wales, the Midlands, Yorkshire, Scotland

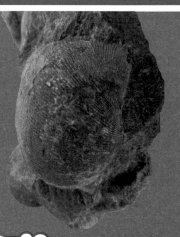

The largest productid brachiopods grew to be over 30 cm wide.

Terebratulid

Age Jurassic and Cretaceous

Appearance Generally smooth spherical shell, like a Roman lamp with a small hole in the larger valve

Where to spot it Jurassic Coast, Cotswolds, Dorset

Rhynchonellid

Age Mainly Jurassic and Cretaceous

Appearance Small, triangular ribbed shells with a wavy edge to both valves

Where to spot them Jurassic Coast, Cotswolds, Norfolk coast

Bivalves

Bivalves are molluscs (animals with soft bodies that have hard shells to protect them) such as clams, oysters and mussels. They lived in fresh and sea water.

Pecten

Age Jurassic to the present day

Appearance Fan-shaped, fairly flat shell with 'ears' along the hinge. Shells are not quite equal in size. Scallops are a modern example

Where to spot it England, south Wales, western Scotland

The logo of a well-known oil company is based upon this shell.

10 POINTS

Trigonia

Age Jurassic

Appearance Medium sized with equal-sized hinged shells; triangle-shaped outline and ridged surface

Where to spot it Dorset, the Midlands

20 POINTS

Venericor

Age Cenozoic

Appearance Medium-to-large size cockle-like shell with two equal-sized valves (hinged shells)

Where to spot it southeast England (especially Bracklesham)

10 POINTS

Gryphea

Age Jurassic

Appearance A primitive oyster with hinged valves of different sizes; the lower shell is large and strongly curved, the smaller one acts like a lid

Where to spot it Dorset, the Midlands, north Lincolnshire, Yorkshire, west coast of Scotland

Gryphea are sometimes known as 'devil's toenails' and is used on Scunthorpe's town crest.

10 POINTS

Gastropods

Gastropods are winkle-like molluscs. They usually have a coiled spiral or conical (cone-shaped) shell.

Turritella

Age Jurassic to the present day

Appearance Long, slender spiral shell; 3–15 cm long

Where to spot it southern England

30 POINTS

Natica

Age Jurassic to the present day

Appearance Coiled, medium-sized shell with a broad cone-shaped appearance; the opening of the shell is oval or rounded; generally smooth

Where to spot it southern England

10 POINTS

Cephalopods

Includes several groups of shelled molluscs as well as squids and octopi. The shells had gas-filled chambers, to stop them from sinking.

Orthoceras

20 POINTS

Age Ordovician to Permian

Appearance A straight, conical shell

Where to spot it Found in the UK in Ordovician and Silurian in particular. Imported Devonian examples are very common as polished ornaments

Goniatite

Age Upper Devonian to Permian

Appearance Coiled shell with outer whorl in contact with the inner whorls. Simple suture pattern

Where to spot it Northumberland coast, Wales, Yorkshire, Scotland

30 POINTS

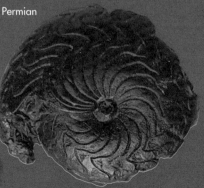

Ammonites

An order of coiled cephalopods which lived during the Jurassic and Cretaceous periods. Their shells are very decorative. They became extinct at the end of the Cretaceous.

Dactylioceras

Age Lower Jurassic

Appearance Coiled shell with open appearance of whorls, 5–10 cm in diameter

Where to spot it Dorset to Yorkshire coast

Used on the town crest of Whitby, where Saxon Saint Hilda is said to have turned a plague of snakes to stone.

20 POINTS

Arnioceras

Age Lower Jurassic

Appearance Coiled shell with strong ribbing and a sharp outer keel

Where to spot it Dorset (the Lyme Regis ammonite pavement), Yorkshire, Isle of Skye

20 POINTS

Hoplites

Age Lower Cretaceous

Appearance Compressed shell, strong ribs and a groove along the outer edge

Where to spot it Southeast England, particularly Folkestone in Kent, where the original brightly coloured shell is preserved

30 POINTS

Puzosia

Age Upper Cretaceous

Appearance Giant, fairly smooth shell with faint ribs

Where to spot it Where chalk crops out: southern England, Norfolk to Yorkshire

Some species grew over 2 m in diameter.

30 POINTS

Belemnites

These are part of the internal skeleton of an extinct squid cephalopod, made of radiating calcite crystals with a conical hole in the blunt end. They lived during the Jurassic and Cretaceous periods.

Cylindroteuthis

Age Upper Jurassic

Appearance long bullet shape; up to 20 cm long

Where to spot it Dorset, the Midlands; small pieces can be common in river gravel

25 POINTS

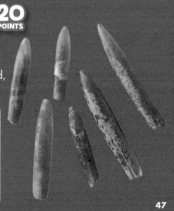

Neohibolites

Age Lower Cretaceous

20 POINTS

Appearance The shape of pointed bullets; up to 5 cm long

Where to spot it Southern England, Yorkshire

Belemnites take their name from the Greek word 'belemnon' meaning dart or javelin. The fossilised rostra (at the back of the belemnites) were believed to be thrown down as darts from heaven during thunderstorms (thunderbolts).

Echinoderms

These fossils include echinoids (sea urchins), crinoids (sea lilies) and rarely, starfish. They are still common in oceans across the world. Crinoids have lived from the Mid Cambrian to the present day.

Pentacrinites

Age Lower to Mid Jurassic

Appearance Crinoid with long, branching arms with a five-sided star-like cross-section; can be 20–40 cm tall

Where to spot it Dorset, Yorkshire

20 POINTS

Amphoracrinus

30 POINTS

Age Silurian to Permian

Appearance Crinoid with long, circular cross-section; often as separate ossicles (St Cuthbert's Beads) or a cast (Screwstone)

Where to spot it Southwest England, northeast England, Pennines

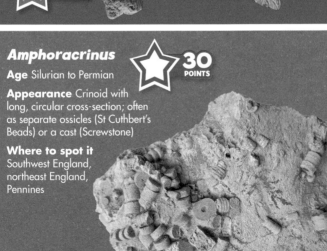

Cidaris

Age Triassic to Holocene

Appearance Circular, like a donut, with pronounced round bosses alternating with five smoother bands and huge pen- or club-like spines, normally found separately

Where to spot it Commonly found as impressions in Cretaceous flint; southeast England, Yorkshire to Norfolk, Chilterns, North and South Downs

40 POINTS

Micraster

Age Upper Cretaceous

Appearance Heart-shaped sea urchin, often preserved as a flint cast

Where to spot it Also commonly found as impressions in Cretaceous flint; southeast England, Yorkshire to Norfolk, Chilterns, North and South Downs

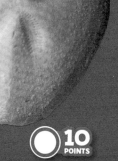

Micraster were called 'fairy loaves' in folklore, and it was believed a household would always have bread if they had one.

10 POINTS

Trilobites

Trilobite bodies were divided into three parts: head, body and tail, and they had a hard cover on their back. They looked like woodlice and some could roll into a ball for protection, but they were more closely related to spiders and scorpions.

Paradoxides

Age Cambrian

Appearance Long oval shape with lateral spines

Where to spot it Pembrokeshire in Wales

 50 POINTS

 TOP SPOT!

Calymene

Age Silurian and Devonian

Appearance About 10 cm long; short, rounded tail

Where to spot it Dudley, the Lake District, Scotland

Calymene is used on the town crest of Dudley, in the West Midlands. Locals call it the 'Dudley Bug'.

30 POINTS

Crabs and lobsters

Crabs, small lobsters and shrimps are frequent discoveries in Jurassic to recent sediments.

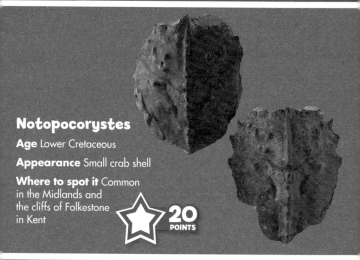

Notopocorystes

Age Lower Cretaceous

Appearance Small crab shell

Where to spot it Common in the Midlands and the cliffs of Folkestone in Kent

20 POINTS

Hoploparia

Age Eocene

Appearance small lobster with long pincers

Where to spot it Essex and Kent

30 POINTS

Insects

Insects are quite unusual finds, but do occur, especially in Lower Cretaceous of southeast England and the arboniferous coal measures of Britain.

Beetle

20 POINTS

Age Lower Cretaceous

Appearance Small (1cm - 2cm), elongated wing covers showing some faint linear ornament

Where to spot them Surrey (similar, but Palaeogene insects in Hampshire and Isle of Wight)

Flies in amber

Flies preserved in Eocene amber are well known and have been picked up on the east coast beaches.

40 POINTS

Fish

Fish first appeared in the Late Cambrian but are very rare as fossils. By the Devonian period, they became more common, with evolved jaws and heavy scales.

Osteolepis

Age Devonian

Appearance Thick, shiny scales on a long, thin fish body with fins and an uneven tail

Where to spot them Orkney, northeast Scotland

TOP SPOT!

This was one of the first fish with a jaw and teeth.

50 POINTS

Shark teeth

Age Silurian to present day

Appearance The main part of the tooth (crown) is usually sharp and pointed, with a root along its base

Where to spot them Kent, Sussex, widely across UK, particuarly Kent and Sussex

The oldest discovered shark fossils are more than 400 million years old! It is very rare to find whole sharks because their skeletons don't tend to become fossils (they are made of cartilage which is much softer than bone).

10 POINTS

Reptiles

Dinosaur bone and tooth

Age Jurassic and Cretaceous

Appearance Dinosaur bones are usually black or brown, with a honeycomb structure on the inside, that is visible when broken

Where to spot them Isle of Wight, Sussex

30 POINTS

Marine reptiles

These include sea-living reptiles like Ichthyosaurs and Plesiosaurs. Bones are quite common, particularly vertebrae (from the spine)

50 POINTS

TOP SPOT!

Age Jurassic and Cretaceous

Appearance Vertebra: concave disc (Ichthyosaur) or more cylindrical (Plesiosaur)

Where to spot them Dorset to Yorkshire

Mammals

Mammoth limb and molar

Mammal fossils are quite unusual, except in Quaternary (Ice Age) deposits where mammoth, deer, bison and other fairly modern types can be found, mainly in river gravel excavations all over the UK.

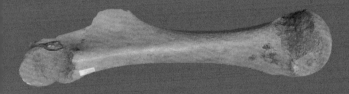

20 POINTS

Other fossils

Graptolites

An extinct animal that lived in colonies from the Middle Cambrian to the Lower Carboniferous. Often found in black shales in Wales, the Lake District and Scotland.

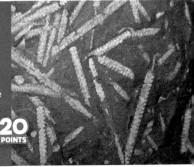

20 POINTS

Worms

Although soft-bodied worms are almost impossible to find fossilised, some, like Serpulids, lived in calcite tubes, which are commonly preserved.

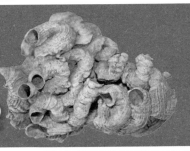

20 POINTS

Forams and Ostracods

Forams are single-celled animals about the size of a pinhead although an exception is *Nummulites*, which can be more than 3 cm across. Wash clay through a fine sieve or muslin to separate the forams and look through a low-power microscope for a whole new world of fossils.

Ostracods are fossil water fleas and can be prepared the same way as forams.

15 POINTS

Trace fossils

These are evidence of plants and animals, left in the sediments that form rock. These can include footprints, burrows, creeping traces, borings in shell or rock, feeding marks and coprolites (fossil poo).

Footprints

Not as uncommon as you might think, dinosaur prints can be seen on the Isle of Wight, the Isle of Skye, in Yorkshire (such as North Bay in Scarborough) and Oxfordshire. There are primitive arthropod trackways in south Wales and Northumberland.

50 POINTS

TOP SPOT!

Burrows

30 POINTS

Many creatures excavated or lived in burrows. Shrimps, sea urchins and bivalves are sometimes found with them. Burrows can also be the only evidence of abundant soft-bodied life, such as worms, in past environments.

Borings

15 POINTS

Wood, shells and rocks can be bored by bivalves, gastropods and sponges.

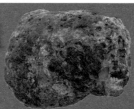

Living fossils

Most of the fossils you will find are from types of animals and plants which are extinct. In some rare cases, modern descendants have managed to survive across the millions of years to the present day.

Ginkgo

Common in Jurassic landscapes, there is only one type of Gingko tree left now (the Maidenhair Tree).

30 POINTS

Horsetail

Compare a modern horsetail with *Calamites* (page 35), its Carboniferous ancestor.

20 POINTS

Coelacanth

A primitive fish believed extinct since the Cretaceous, it has now been found living in the Indian Ocean, though it is rare. Look for a preserved coelacanth in a UK museum (score double points for a fossilised one).

TOP SPOT!

50 POINTS

Lingula

A small brachiopod which has been present on Earth for 500 million years.

30 POINTS

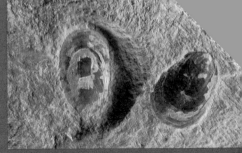

R✺CKWATCH

What is Rockwatch?

Rockwatch is the junior club of the Geologists' Association, a UK society for geologists and earth scientists. If you want to understand and explore Earth and its processes, this is the club for you!

How can I become a member and what is included?

Joining Rockwatch is easy. A grown-up can complete the membership form online and you will receive your membership pack in a few days.

The pack contains everything a young geologist needs, including a magnifier, fact cards, a grain size card and a geological map of Britain and Ireland.

You will also receive a magazine three times a year, which contains expert articles, activities, puzzles and fact cards to add to your collection.

Rockwatch members get invited to special events and field trips in different places around the UK, including working quarries and beaches of special geological interest. You will have the chance to collect minerals and fossils with other Rockwatch members and meet knowledgeable experts.

Last but not least, as a member you will get the chance to enter the exciting Rockstar competition every year with the chance to win amazing prizes! Here are some prize-winning craft entries.

How can I find out more?

For more information, see www.rockwatch.org.uk. You can contact Rockwatch via the contact form on the website, or by email at hello@rockwatch.org.uk.

Glossary

boss circular, raised ornamentation

composition what a rock or mineral is made of

crystal the natural shape of a mineral when it can form freely; a crystal has flat surfaces (faces) arranged in a regular and repeating pattern. In igneous and metamorphic rocks, the grains are still known as crystals

geology the study of the earth: how it works, what it is made of, and past life

granular made of grains of roughly the same size; can apply to any rock type

intruded used to describe very hot, melted igneous rock which has forcefully moved into another rock, before cooling to become solid itself

keel a ridge, running round the edge of something

lustre used to describe a mineral's sheen (how shiny or not it is); special words include metallic (looks like metal), glassy, pearly (like a pearl), or dull

massive used to describe minerals with no obvious crystal shape, and in rocks where there are no obvious structures (e.g. layers)

nodule a small, rounded lump

ore a rock from which one or more valuable minerals can be removed; minerals containing metals are very important

ossicle a small part of a larger skeleton

resinous has a lustre like wax, glue, chewing gum or resin

rhombohedral describes a six-sided object whose opposite sides are parallel, like a box that has been pushed over in two directions

rosette shaped liked a rose flower, or formed of groups of petal-shaped crystals

shear surfaces where slices of rock have slipped against other rock surfaces

streak the colour of a powdered mineral. It can be found by making a small mark with the mineral on the back of a porcelain kitchen tile

suture the junction between two separate parts of a shell or bone

tabular flattened shape in minerals, like a table top

vein a sheet-like rock made of minerals which have filled in a fracture (break) in another rock

whorl a circular or spiral shape

Index

I-SPY How to get your i-SPY certificate and badge

Let us know when you've become a super-spotter with 1000 points and we'll send you a special certificate and badge!

Here's what to do:

- Ask an adult to check your score.

- Apply for your certificate at www.collins.co.uk/i-SPY (if you are under the age of 13 we'll need a parent or guardian to do this).

- We'll email your certificate and post you a brilliant badge!

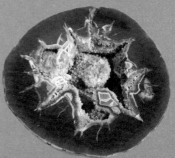